BALLAST WATER TREATMENT SYSTEMS

BALLAST WATER

A GUIDE ON BALLAST WATER TREATMENT SYSTEM

WRITTEN BY
SUNIL SERANGI

BALLAST WATER TREATMENT SYSTEMS

Sunil Sarangi is an experienced Marine Engineer, He has founded Eco Marine, a Ballast Water Treatment Systems Design and consultants' firm that helps to protect and control aquatic invasive species.

He began working in the Ballast water regulation in the United States in the early 2005's. He received his bachelor's degree in Marine Engineering from the Birla Institute of Technology at India in 2004. He also, received his Master's of Science in Ocean Engineering from Florida Institute of Technology at Florida in 2008. As his career continued having a top stop, he bagged Master of Science in Ocean Engineering with a Bachelor of Science in Marine Engineering, and also MBA.

During the past decade his work has focused upon developing cost-effective methods for reducing the spread of aquatic invasive species, the economic cost of this introduction has been estimated by the U.S. Fish and Wildlife Service at about $5 billion.

His work over the past 10 years can be summarized as covering the following subjects:

- The design and development of Ballast Water Treatment Systems
- Investigations into the effectiveness of techniques and technologies for measuring and reducing aquatic nuisance species.
- As a lead manager, successfully completed more than $40 million USD projects related to Ballast Water Treatment Systems.

Mr. Sunil Sarangi has authored several articles on Marine engineering and Ballast Water Treatment issues and has spoken at numerous symposia and conferences during the past decade.

Contents

CHAPTER 1 4

BALLAST WATER: INTRODUCTION 4

A key objective 7
Significance of ballast water 15
Some management options 20
Policy and Legal Framework for Ballast Water Management 21
CHAPTER 2 23

GLOBAL LEGAL FRAMEWORKS ADDRESSING SPECIES INTRODUCTION ... 23

Legislation on Ballast Water 23
Ballast Water Management Under the Ballest Water Mnangement Convention 26
The Ballast Water Management Convention 27
What Is Ballast Water Management? 28
Ballast Water Management Requirements 30
Ballast Water Exchange Standard: D-1 Standard 31
CHAPTER 3 33

BALLAST WATER MANAGEMENT SYSTEMS FOR VESSELS 33

Introduction 34
IMO Requirements 35
Key Features of Eco Guardian – Ballest's Water Management System (BMWS) 36
System General 37
During voyage 39
CHAPTER 4 44

BALLAST WATER TREATMENT SYSTEMS 44

Treatment 45
Results 46
Comparison of BWTSs using UV radiation 46
UV dose response experiment 47
Comparison of chlorine-based treatment systems 47
EC dosage experiment 48
Comparison of UV and EC treatment systems 48
Comparison of UV BTWS 49
UV dosage 50
Comparison of CD and EC BWTS 50
EC Dosage 51
Comparison of UV and CD/EC BWTS 51
Protecting Aquatic Ecosystems 56
Why is ballast water treatment needed? 61
CONCLUSION 64

CHAPTER 1
BALLAST WATER: INTRODUCTION

The human-mediated introduction of species to regions of the world they could never reach by natural means has had great impacts on the environment, the economy, and society. In the ocean, these invasions have long been mediated by the uptake and subsequent release of ballast water in ocean-going vessels. Increasing world trade and a concomitantly growing global shipping fleet composed of larger and faster vessels, combined with a series of prominent ballast-mediated invasions over the past two decades, have prompted active national and international interest in ballast water management. Assessing the relationship between propagule pressure and invasion risk in ballast water informs the regulation of ballast water by helping the Environmental Protection Agency (EPA) and the U.S. Coast Guard (USCG) better understand the relationship between the concentration of living organisms in ballast water discharges and the probability of nonindigenous organisms successfully establishing populations in U.S. waters. The report evaluates the risk-release relationship in the context of differing environmental and ecological conditions, including estuarine and freshwater systems as well as the waters of the three-mile territorial sea. It recommends how various approaches can be used by regulatory agencies to best inform risk management decisions on the allowable concentrations of living organisms in discharged ballast water in order to safeguard against the establishment of new aquatic nonindigenous species, and to protect and preserve existing indigenous populations of fish, shellfish, and wildlife and other beneficial uses of the nation's waters. Assessing the relationship between propagule pressure and invasion risk in ballast water provides valuable information that can be used by federal agencies, such as the EPA, policy makers, environmental scientists, and researchers.

It is widely accepted that more than 90 % of cargoes in international trade are safely transported by ships throughout the world, and the carriage of ballast water plays an essential role in guaranteeing the safe navigation and operation of such ships. At the same time, though, ballast water poses an environmental threat by serving as a vehicle to transport live unwanted species across the oceans. According to different estimates, up to 10 billion tonnes of ballast water is transported around the world by ships annually, and several thousands of microbial, plant and animal species may be carried globally in ballast water. When these species are discharged into new environments, they may become established and can also turn invasive, thus severely disrupting the receiving environments with the potential to harm human health and the local economy. The global economic impacts of invasive marine species are difficult to quantify in monetary terms, but are likely to be of the order of tens of billions of US dollars per year. Consequently, the introduction of harmful aquatic organisms and pathogens to new environments, including via ships' ballast water, has been identified as one of the four greatest anthropogenic threats to the world's oceans. The International

Maritime Organization (IMO), the United Nations' specialized agency responsible for the safety and security of shipping and the prevention of marine pollution by ships, first responded to this issue by developing guidelines and recommendations aimed at minimizing the transfer of live organisms

and pathogens by exchanging ballast water at sea, since experience had shown that ballast water exchange in deep waters reduces the risk of species transfers. At the same time, it was recognized that higher levels of protection could be reached with other protective measures, e.g. through ballast water treatment. It also became clear at the time that a self-standing international legal instrument for the regulation of ballast water management would be necessary to avoid regulatory action by authorities at national, provincial and even local levels.

This could have resulted a fragmented, patchwork-like ballast water management approach which had to be avoided by all possible means in an eminently cross-border vi industry like shipping. Consequently, IMO developed the globally applicable International Convention for the Control and Management of Ships' Ballast Water and Sediments (BWM Convention), which was adopted in February 2004 at a diplomatic conference in London.

This instrument will enter into force 12 months after the date on which more than 30 states, with combined merchant fleets not less than 35 % of the gross tonnage of the world's merchant shipping, have ratified it. As of December 2013, 38 states representing 30.38 % of the world merchant shipping gross tonnage had ratified the BWM Convention. IMO has also joined forces with the Global Environment Facility (GEF) and the United Nations Development Programme (UNDP) to implement the Global Ballast Water Management Programme (GloBallast), which was followed by the GloBallast Partnerships Programme.

A key objective

A key objective of these programmes is to provide assistance, mainly to developing countries, for the implementation of the BWM Convention. The BWM Convention introduces new requirements for port States and ships all around the world, although its implementation is a complex process. Despite the global efforts of industry, member states and IMO over many years, efficient, economically feasible, environmentally friendly and safe methods of preventing the translocation of harmful organisms via ballast water are still being developed. The implementation of some of the ballast water management methods becomes even more complicated due to the difficulties encountered in their applicability because of the differences in shipping patterns and geographical specifics. The shipping industry on one side and coastal states on the other are confronted with serious obstacles when trying to find simple solutions to the extent that turnkey solutions may need to be developed on a case-by-case or port-by-port basis, this without causing an excessive burden to the shipping industry and, consequently, to the global trade. With great interest and appreciation, I note that this book summarizes comprehensively the current knowledge regarding the multifaceted ballast water issue. It provides an overview of the possible solutions to the complex issue of ballast water management and also outlines consequences and implications to address the ballast water "problem" following the provisions of the BWM Convention. It delivers an excellent overview regarding ships' ballast operations; environmental and other aspects of the issue; and international requirements as well as an in-depth analysis of possible ways to approach or manage the challenge in the most effective way.

I am convinced that this book will be an invaluable tool for you and those that are interested in marine environment protection and, most of all, will provide much needed assistance to maritime administrations when trying to ratify and implement the BWM Convention. Motril , Spain Miguel Palomares December 2013 Former Director of the IMO Marine Environment Division Foreword vii Foreword The rapid growth of global economic trade and the seemingly unlimited human mobility around the world, commencing in the mid-1800s, opened many windows of opportunity for trading goods not only between population centers but also into remote places of the world. In the twenty-first century, transportation by trucks, trains and planes is surpassed by far in volume and distance travelled by the shipping and boating industries – trans- and inter-oceanically via container ships, bulk carriers, and tankers and coastally by both cargo vessels and a vast fleet of recreational and fishing vessels. It thus does not come as a surprise that the issue of unintentional transmission of organisms (including pathogens) across oceans and continents has reached a new dimension that is of serious concern to maintain and sustain ecosystem integrity and ecosystem services.

Aquaculturists in coastal and marine waters have been aware of the problems of transfers of exotic species since the end of World War II, being especially affected by the unintentional introductions of fouling organisms and disease agents. While the aquaculture industry was often blamed for self-contamination (which was certainly a valid point and partially true with disastrous examples), we know today that many of the problems with exotic fouling organisms affecting aquaculture and other

stakeholders also originated from the shipping industry through the long-term uncontrolled release of ballast water and transfer via hull, sea chest, and other fouling. Aquatic biodiversity and environmental health have been on the agenda of ecologists for decades.

Most concern has been expressed for the potential of "loss of biodiversity " in light of increasing anthropogenic pressures. This concern has been expressed by many organizations, while national and international regulatory authorities try to include biodiversity issues into environmental management schemes. However, early on in the biodiversity debate, fewer scientists pointed to the fact that we are not only dealing with the " loss of biodiversity " but also with a " change " or " increase " of species diversity due to human intervention and that these changes may also be considered as threats to ecosystem stability and services.

Thus, some recent literature has argued that adding species to natural communities viii is beneficial, but these arguments typically do not address the fundamental changes that accompany such additions, such as the often vast decrease in the abundance of native species (even if these still remain, somewhere) and the concomitant cascades in altering energy flow, competition, and predator–prey relationships. Australia, New Zealand, the United States and Canada provided pioneering research work in the area of marine bioinvasions and ballast water by delineating the dimensions of the problem commencing in the 1970s and 1980s.

In Europe and the rest of the world, studies on the dimension of the problem started at least a decade later. Commencing in the 1990s, international conventions and organizations (such as the United Nations' International Maritime Organization (IMO), responsible for the safety and security of global shipping and the prevention of marine pollution by ships) began to be concerned about and involved in the promulgation of regulatory frameworks to minimize the risks associated with the increasingly huge volumes of ballast water transfer and biofouling on commercial and recreational vessels.

Similarly, over the past two decades, national regulatory frameworks have been developed in a number of countries. All of these management scenarios, however, depend on sound and solid research results to properly and effectively reduce the risk of transfer of (potentially) harmful organisms. The authors of this book are among the pioneers who intensively studied the role of shipping and have been at the forefront (in cooperation with others worldwide) to promote the development of methods on how to (a) monitor the fate of nonindigenous species transferred by ballast water,

(b) Standardize mitigation and control procedures for practical application by industry and regulatory authorities, and

(c) Develop the much-needed risk assessment and "hotspots" identification where protective action is needed most. Their work, together with many other scientists and organizations, contributed to the preparation of the International Convention for the Control and

Management of Ships' Ballast Water and Sediments, adopted by IMO in 2004. This book is very timely, providing a comprehensive state-of-the-art synthesis: during the past two decades, tremendous progress had been made in research to understand both the importance of these transmission vectors and the environmental risks associated with them. The authors have contributed greatly both through original research and practical testing and extensive review work to our present knowledge on mitigation strategies and treatment procedures. The present volume builds and expands on previous overviews where the authors have been instrumental in providing concepts and guidance to help developing solutions to the problem.

The undersigned, having been involved in cooperative work with the authors over many years, are pleased to see this progress reported and summarized in a format that will not only be of great value to experts in the field but also provide both the background and the current state of knowledge to a much broader audience interested in issues related to the unintentional global transfer of species. The engagement of a wide audience via this book's modern and practical summary of Foreword ix global ballast water management will assist greatly in encouraging all stakeholders to more vigorously implement the required management schemes that will reduce invasions and thus their impact on our environment and economy.

Today global shipping transports over 90 % of the world's overseas trade and trends anticipate that it will continue to play an increasing role world-wide. Shipping operations inevitably include also pressures on natural environments. The most recent waterborne threat is the transfer of harmful aquatic organisms and pathogens with ballast water and sediments releases, which may result in harmful effects on the natural environment, human health, property and resources globally. The significance of the ballast water issue was already addressed in 1973 by the International Maritime Organization (IMO) as the United Nations specialised agency for the regulation of international maritime transport at the global scale. Committed work by many experts, scientists, politicians, IGOs and NGOs at IMO resulted in the adoption if the International Convention for the Control and Management of Ships ' Ballast Water and Sediments (BWM Convention) in February 2004, which is now to be ratified and implemented.

Work on ballast water management issues has also shown to be very complex, hence there are no simple solutions. Nevertheless, the BWM Convention represents a globally uniform framework for the implementation of ballast water management measures, and different supporting tools like risk assessment and decision support systems have been developed to support its efficiency. In this chapter the reader is introduced to various ballast water issues and responses to it. The intention of this book and the overview of its content is also presented.

General Introduction The continuous intensification of the globalization of trade and production increased the demand for new, faster and more frequent linkages among trading and commodity production areas. These transport demands can only be met by maritime shipping because of its inherent technical and technological advantages and properties. The shipping industry has reacted to these needs with new and more frequent connections, increased vessels cargo and passenger capacity, and new vessel types and technologies.

Today global shipping transports over 90 % of the world's overseas trade. Future trends anticipate that global and local shipping play an increasing role world-wide. Intensified shipping and related developments has also resulted in disasters of unprecedented dimensions. Widely known examples include the Titanic in 1912, Torrey Canyon in 1967, Amoco Cadiz in 1978, Exxon Valdez in 1989, Estonia 1994, Sea Empress in 1996, Erika in 1999, and Prestige in 2003.

Such disasters resulted in the loss of human lives, property and/or caused significant damage to coastal ecosystems. In addition another inevitable consequence of shipping disasters is the pollution of the environment caused by a variety of pollutants. Apart from harmful effects as consequences of shipping disasters, regular shipping activities cause other negative environment effects, e.g., sea pollution through the discharges of oily water and sewage from vessels, air pollution from exhaust gases emitted by the vessel's machinery, pollution of water and marine organisms by toxic protective underwater hull coatings (anti-fouling paints), and one of the most recent waterborne threats – the transfer of harmful aquatic organisms and pathogens (HAOP) with ballast water and sediments releases.

Given its 'mysterious' nature in combination with severely harmful effects on the natural environment, human health and the global economy, the problem has attracted attention of scientists and the public worldwide, which was particularly advanced in the 1980s and 1990s due to severe impacts of only a few introduced species. What is the problem? Vessels need additional weight as a precondition for safe navigation in cases when they are not carrying cargo or are not fully or equally laden.

The weight adding material is referred to as ballast. Historically, ballast was solid (e.g., sand, rocks, cobble, iron). With the introduction of iron, replacing wood, as basic vessel building material in the middle of the nineteenth century, the doors are opened to new ballasting technologies. Loading of water (i.e., ballast water) in cargo holds or ballast water tanks has shown to be easier and more time efficient compared to solid ballast. Therefore, water as ballast was adopted as a new practice of increasing importance. Many different types of vessels have different structures of ballast tanks, as well as different ballast system capacities. Vessels ballast water operations are related to vessel type, vessel construction, cargo operations and weather conditions.

However, there are no clear limits among all these factors, but the decision on ballast water operations is under the discretion of the chief officer and direct control of the captain, who is responsible for the vessels stability and safety. Nowadays vessels fundamentally rely on ballast water for safe operations. A model for the assessment of ballast water discharges has been developed and tested. It is estimated that global ballast water discharges from vessels engaged in the international seaborne trade in 2013 would be approximately 3.1 billion tonnes.

Significance of ballast water

The significance of the ballast water issue was already addressed in a 1973 International Maritime Organization (IMO) Resolution (IMO 1973). IMO as the United Nations specialized agency for the regulation of international maritime transport at the global scale, was tasked to deal with this issue further. After more than one decade of intensive and committed work by many experts, scientists, politicians, IGOs and NGOs at IMO, the final text of the International Convention for the Control and Management of Ships' Ballast Water and Sediments (BWM Convention) was completed and adopted in February 2004 at a diplomatic conference in. The BWM Convention introduced new BWM related requirements for port States and vessels all around the world. However, the implementation of this Convention is far from being simple. After the adoption of the BWM Convention several countries and regions have implemented (voluntary) ballast water management approaches.

Due to global efforts of industry, Member states and IMO, efficient, financially feasible, environmentally friendly and safe methods of preventing the translocation of HAOP via ballast water are developed.

More than 30 ballast water management systems (BWMS) have already been certified (type approved) so that most vessels can today be equipped with such systems. We are aware that this is a very fast developing area and market, at least 20 more systems are currently in the certification process. The BWM Convention is at the moment of this writing not yet in force, but does today represent a solid and uniform framework for preventive measures to avoid HAOP introductions and it needs to be implemented by individual countries or joint approaches. The BWM Convention enters into force 12 months after the date on which more than 30 states, with combined merchant fleets not less than 35 % of the gross tonnage of the world's merchant shipping, have signed this Convention. As of December 2013, 38 states ratified the BWM Convention, representing 30.38 % the world merchant shipping gross tonnage.

Nonetheless it must be emphasized that efficient ballast water management (BWM) does not imply the prevention of HAOP introductions at any cost, thereby laying an additional burden on and generating higher costs for the shipping industry. Undoubtedly, the cost of prevention should not be higher than the benefits it generates. Conditioned by the lack of on board installed BWMS on existing vessels, ballast water exchange (BWE) is today the most widespread available BWM method also approved by the BWM Convention.

Nevertheless, ballast water exchange has drawbacks which make it inefficient or even impracticable under certain conditions (e.g., on shorter voyages where "intended routes" are too close to the shore, attain insufficient water depths, a lack of knowledge of the presence of HAOP in the water exchange area). Further, other issues related to an efficient BWM system arise which are outside of the vessels' responsibility, e.g., targeting of vessels for ballast water sampling as part of port State compliance control procedures. As a result, countries wishing to protect their seas, human health, property and resource from the introduction of HAOP with ballast water are confronted with a significant challenge.

Given that BWM requirements may result in inefficiencies, lower safety margins and higher costs in the shipping industry, the reasons described above make the 'blanket approach' (i.e., mandatory BWM for all ships) unjustifiable in a range of different local conditions.

An alternative to the blanket approach is the 'selective approach' where BWM is required for selected vessels. This selection should be based on a suite of information needs and procedural decisions to aid transparent and robust BWM decisions. Such systems have been developed in a variety of applications where a large number of complex decisions must be made in a consistent, transparent and defensible manner.

These systems are typically referred to as decision support systems (DSS). Such a DSS as applied to BWM implies adjusting the intensity level of BWM measures to each voyage based on risk assessment (RA), and recommends also compliance monitoring and enforcement (CME) actions.

A BWM DSS provides essentially needed support to responsible agencies for the implementation of effective BWM measures. The introduction of BWM practices adds burden and costs mostly to the shipping industry, on the other side, their efficiency is critical. In light of these, the BWM DSS needs to provide for:

– an effective protection against the introduction of HAOP;

– proper RA as one of the key elements of the BWM DSS;

– local specifics are addressed in direct relation with the effectiveness of the BWM (e.g., geographical, hydrological, meteorological, important resources, shipping patterns, regulatory regime);

- a selection of most effective and safe BWM methods according to the RA;
- the consideration of impacts to the shipping industry (including safety);
- the consideration of impacts on international trade;
- timely decision making;
- the reduction of subjectiveness in the decision process; and
- a consistent and transparent decision making process.

A uniform DSS methodology and RA concerning HAOP introductions via ballast water has not yet been developed. Several foundations have already been laid, e.g., Australian DSS, GloBallast Ballast Water Risk

Assessment Environmental Ballast Water Management Assessment – EMBLA and BWM RA and DSS for Slovenia. More recently BWRA according to the BWM Convention requirements was developed for

HELCOM and OSPAR. Currently BWRA and BWM DSS for European Seas is being developed under the EU-funded VECTORS project, and for the Adriatic Sea under the IPA Adriatic strategic project BALMAS. Yet the complexity and intrinsically modern character of the problem leaves several questions, as revealed by the inefficiency of these applied systems, unansared.

The need for answers bears vital significance for the international environment, the goal being the future implementation of an efficient BWM system in tandem with considerations for a sustainable shipping industry.

Commercial vessels are built for the transport of various cargoes or passengers. When a vessel is not fully laden, additional weight is required to pro- vide for the vessel's seaworthiness, e.g. to compensate for the increased buoyancy which can result in the lack of propeller immersion, inadequate transversal and longitudinal inclination, and other stresses on the vessel's hull. The material used for adding weight to the vessel is referred to as ballast. Historically, ballast material was solid, but after the introduction of iron as basic vessel building material in the middle of the nineteenth century, loading of water (i.e., ballast water) in cargo holds or tanks had shown to be easier and more efficient. Even when a vessel is fully laden it can require ballast water operations due to a non-equal distribution of weights on the vessel, weather and sea conditions, an approach to shallow waters, and the consumption of fuel during the voyage.

As a result of these factors, vessels fundamentally rely on ballast water for safe operations as a function of their design and construction. This chapter describes vessel's ballast water systems, ballast tank designs, ballasting and deballasting processes as well as safety and legislative aspects of ballast water operations. In addition a detailed ballast water discharge assessment model is provided. Using concepts of this model an estimation of global ballast water discharges from vessels engaged in the international seaborne trade was estimated as 3.1 billion tonnes in 2013.

Some management options

There is a wealth of policy and management options addressing species introductions including conventions, treaties, multilateral agreements and codes of practices. Together these instruments support an internationally consistent management of specific transport vectors, quarantine or other biosecurity measures. This chapter lists selected global legal frameworks addressing species introductions. Chronologically, the first international instrument on unintentional introductions may have been the International Health Regulations issued in 1969 by the World Health Organization (WHO). These regulations are prepared to support public health care operations and to ensure the prevention of the spread of epidemics (e.g. plague, cholera). This chapter addresses legal frameworks addressing species introductions with the focus on ballast water related policy and legal frameworks. It gives an update on the current status of ballast water management requirements world-wide. A number of countries have taken the approach to nationally imple- ment ballast water management requirements.

We describe that most of these national requirements are based upon the IMO Ballast Water Exchange Standard, some countries refer to the Ballast Water Performance Standard and a minority addresses land-based ballast water reception facilities.

Policy and Legal Framework for Ballast Water Management

There is a wealth of policy and management options to combat the introduction of species including conventions, treaties, multilateral agreements and codes of practices which aim to support an internationally consistent management of specific transport vectors, quarantine or other biosecurity measures. These instruments regulate species transfers, control their release or address mitigation measures for introduced species populations by, e.g., eradication programmes. This chapter focusses on ballast water related policy and legal frameworks and gives an update on the current status of ballast water management (BWM) requirements world-wide.

The International Convention for the Control and Management of Ships' Ballast Water and Sediments, London 2004 (BWM Convention) is considered as the basic global framework for BWM measures. International and national legislation pro- vide for the prevention of harmful impacts caused by discharges of Harmful Aquatic Organisms and Pathogens (HAOP) via ballast water.

Port States need to prevent unlawful acts of vessels flying their flag (i.e., Flag state obligations), as well as those occurring in their jurisdictional waters (i.e., Port State obligations) (IMO 2004).

CHAPTER 2
GLOBAL LEGAL FRAMEWORKS ADDRESSING SPECIES INTRODUCTION

Chronologically, the first international instrument to address unintentional introductions may have been the International Health Regulations issued in 1969 by the World Health Organization (WHO). These regulations are prepared to provide support to public health care operations and to ensure the prevention of the spread of epidemics (e.g., plague, cholera).

Legislation on Ballast Water

The International Maritime Organization (IMO) noted the negative impact of non- indigenous organisms transported in the ballast water of ships as far back as in the early 1970s (IMO 1973a). At the International Conference on Marine Pollution in 1973 the Resolution on the Research into the Effect of Discharge of Ballast Water Containing Bacteria of Epidemic Diseases was adopted.13 After the acknowledgement of the problem in a 1973 resolution, the IMO, through its Marine Environment Protection Committee (MEPC), started to develop an instrument to cope with this problem in the early 1990s (IMO 1993).

The International Convention for the Prevention of Pollution of Ships.14 (MARPOL) adopted by IMO in 1973 contains six regulatory annexes, each relative to a specific source of ship-generated pollution: pollution by oil, pollution by noxious liquid substances, pollution by harmful substances in packaged form, pollution by sewage from ships, pollution by garbage from ships, air pollution from ships (IMO 1973b). Initially the problems concerning ballast water discharge are to be regulated by a new annex for the prevention of uncontrolled ballast water discharge. According to the MARPOL Convention, harmful substances can be defined as any substances dumped into the sea that pose a risk to human health, are noxious to live sea organisms, or disturb any legitimate use of the sea, and should therefore as such be controlled. It was later agreed at IMO that ballast water cannot be viewed as pollutant and could therefore not be covered by MARPOL.

As a first effort, the International Guidelines for Preventing the Introduction of Unwanted Aquatic Organisms and Pathogens from Ships Ballast Water and Sediment Discharges are adopted at the 31st Session of MEPC in July 1991. In 1993, the IMO Assembly adopted these Guidelines by Resolution A.774 (IMO 1993).

It soon became clear thereafter that species' movements in ballast water cannot be completely prevented, and work on this matter continued at IMO.

In 1997, the Guidelines for the Control and Management of Ships' Ballast Water to Minimize the Transfer of Harmful Aquatic Organisms and

Pathogens are adopted by Resolution A.868(20) (IMO 1997), which replaced Resolution A.774(18).

The importance of biological invasions was brought into greater focus as several devastating introductions in many countries occurred (e.g., the Atlantic comb jelly, Mnemiopsis leidyi, in the Black Sea, the zebra mussel, Dreissena polymorpha, in the North American Great Lakes, the Northern Pacific sea star, Asterias amurensis, in Australia and Tasmania, and it was recommended that IMO works towards a stand-alone Convention to address this problem. Consequently the BWM Convention was finalised and adopted at the Diplomatic Conference in London on February 13, 2004.

A number of countries have taken the approach to nationally implement BWM requirements of which some have also ratified the BWM Convention. Most of these requirements are based upon the IMO Ballast Water Exchange (BWE) Standard (Regulation D-1), some countries refer to the Ballast Water Performance Standard (D-2 standard) and a minority addresses land-based ballast water reception facilities. Should BWE not be possible due to, e.g., safety reasons, most countries require that the next port of call should be notified that other measures can be taken, which includes a BWE in a designated coastal area or other water treatment (e.g., brine treatment in Canada). Some countries further request ships to have a BWM plan and an up-to-date ballast water record book on board.

Ballast Water Management Under the Ballest Water Mnangement Convention

The importance of ballast water as a vector for moving non-indigenous species was initially addressed in a 1973 International Maritime Organization (IMO) resolution. Subsequently IMO worked towards the finalization of the International Convention for the Control and Management of Ships' Ballast Water and Sediments (BWM Convention) which was adopted in February 2004 at a diplomatic conference in London. The BWM Convention's main aim is to prevent, minimize and ultimately eliminate the risks to the environment, human health, property and resources which arise from the transfer of harmful aquatic organisms and pathogens via ships' ballast waters and related sediments.

It should be noted that harmful aquatic organisms in this context are not limited to non-indigenous species, but covers all aquatic species irrespective of their origin. As defined at IMO "Ballast Water Management means mechanical, physical, chemical, and biological processes, either singularly or in combination, to remove, render harmless, or avoid the uptake or discharge of Harmful Aquatic Organisms and Pathogens within Ballast Water and Sediments."

The BWM Convention and its supporting guidelines are described in this chapter, outlining the ballast water exchange and performance standards, warnings concerning ballast water uptake in certain areas, ballast water reception facilities, sediment management as well as exemptions and exceptions from ballast water management requirements. This chapter ends with the description of implementation options of the BWM Convention.

The Ballast Water Management Convention

The importance of ballast water as a vector for moving non-indigenous species was initially addressed in a 1973 International Maritime Organization (IMO) resolution (IMO 1973). Subsequently IMO worked towards the finalization of the International Convention for the Control and Management of Ships' Ballast Water and Sediments (BWM Convention) which was adopted in February 2004 at a diplomatic conference in London (IMO 2004). This Convention's aim is to prevent, minimize and ultimately eliminate the risks to the environment, human health, property and resources which arise from the transfer of harmful aquatic organisms and pathogens (HAOP) via ships' ballast waters and related sediments. It should be noted that harmful aquatic organisms in this context are not limited to non-indigenous species, but covers all species irrespective of their origin.

The Regulations for the control and management of ships' ballast water and sediments are presented in five sections:

Section A: General provisions: Definitions, General applicability, Exceptions, Exemptions, Equivalent Compliance;

Section B: Management and control Requirements for Ships: Ballast Water Management;

Section C: Special Requirements in Certain Areas;

Section D: Standards for Ballast Water Management; and

Section E: Survey and Certification requirements for Ballast Water Management.

Certain obligations are to be met by all stakeholders including the ship, the Administrations, i.e., both in their capacity as Flag state, Port State, and as the representative of a Party, and IMO.

What Is Ballast Water Management?

As defined at IMO: "Ballast Water Management means mechanical, physical, chemical, and biological processes, either singularly or in combination, to remove, render harmless, or avoid the uptake or discharge of Harmful Aquatic Organisms and Pathogens within Ballast Water and Sediments."

BWM in its core sense means the prevention, minimization and ultimate elimination of the transfer of HAOP via vessels' ballast waters and sediments. In light of this, BWM cannot only be understood as mechanical, physical, chemical, and bio- logical processes preventing the transfer of HAOP, because the process includes also different precautionary measures to minimize the uptake of HAOP and sediments. Those include the avoidance of ballast water uptake, where practicable,

- in areas identified by the port State in connection with advice provided by ports;

- in darkness when the organism concentration in upper water layers increases;

- in areas with outbreaks, infestations or known populations of HAOPs;

- in very shallow water because it is more likely to pump in bottom living organisms;

- where propellers may stir up sediment;

- where dredging is or recently has been carried out; and

- nearby sewage outfalls.

Furthermore, no mixing of ballast water should occur and additional management practices may apply, e.g., risk assessment (RA) decision support system. Hence BWM should be under- stood as a complex, multi-facetted process of all precautionary measures, preventive and treatment procedures, as well as additional measures taken to prevent, minimize and ultimately eliminate the transfer of HAOP via ballast water and sediments.

Vessels should also, whenever possible, implement precautionary practices, i.e., avoid the unnecessary discharge of ballast water. Should it be necessary to take on and discharge ballast water in the same port to facilitate safe cargo operations, unnecessary discharge of ballast water that has been taken up in another port should be avoided. Managed ballast water which is mixed with unmanaged ballast water is no longer in compliance with Regulations D-1 and D-2.

Ballast Water Management Requirements

By the basic principle, vessels (not ports) are required to conduct BWM according to the requirements of the BWM Convention. However, port reception facilities are also considered by the BWM Convention as a BWM option, i.e., Regulation B-3.6 and Guidelines for ballast water reception facilities (G5) (G5 Guidelines) (IMO 2006b). During the BWM Convention negotiations ballast water reception facilities are considered as the primary BWM measure. However, as ships may need to conduct ballast water operations also outside ports, such reception facilities would not cover all ballast water discharges. Therefore, treatment on board ship before ballast water discharge is required.

Standards for BWM are dealt with by the BWM Convention in Regulations D-1 and D-2. The BWM Convention introduces these two different protective regimes as a sequential implementation regime:

Ballast Water Exchange Standard (Regulation D-1, so called D-1 standard) requiring ships to exchange a minimum of 95% ballast water volume;

Ballast Water Performance Standard (Regulation D-2, so called D-2 standard) requires that the discharge of ballast water have the number of viable organisms below the specified limits.

The D-2 standard is based on a limited number of organisms that can be dis- charged with ballast water. The phase-in of the D-2 standard was originally planned gradually, based on the vessels total ballast tanks capacity and if these vessels are existing or are new builds. When the phase-in dates are set, the expectation was that technology and manufacturing capacity would be first available for vessels with lower ballast water capacities and flow rates. As such dates are set to allow a gradual maturity of the technology with the expectation that the very high flow rates would come later due to the technical challenges. These include that on smaller vessels due to engine room limited space it might be difficult to install ballast water management systems (BWMS) at that time. Higher flow rates are considered difficult as the first generation of BWMS was not able to meet these flow requirements.

However, the BWM Convention has not come into force and certain phase-in dates have already passed. This resulted in a debate at IMO and Marine Environment Protection Committee (MEPC) at its 65th session approved a draft IMO Assembly resolution on the application of Regulation B-3 of the BWM Convention, which addresses the fixed dates, to ease and facilitate the smooth implementation of the BWM Convention. This was approved at the 28th session of the IMO Assembly.

Ballast Water Exchange Standard: D-1 Standard

Approximately 10 years ago when the D-2 standard was negotiated at IMO no BWMS was readily available. In the absence of full scale BWMS to be installed on vessels, it was suggested by MEPC that ballast water exchange (BWE) at sea may reduce the risk of species introductions. Most vessels are enabled to conduct a BWE without needing extra installations.

The reasoning behind BWE is that coastal organisms pumped on board during ballast water uptake, when discharged at sea are unlikely to survive due to, e.g., salinity issues and the lack of a hard substrate to complete their life cycle. In addition, high sea organisms when pumped on board during the BWE will unlikely survive when released in coastal waters also due to possible salinity changes and the lack of suitable habitats. Further, it is well-known that organism concentrations are much lower in high seas compared to coastal waters which reduces the risk of species introductions. However, sampling studies on board of commercial vessels have shown that in certain instances after BWE a higher concentration of organisms was found in the ballast water. This specifically occurred when the BWE was undertaken in shallower seas or during high organism concentrations, such as algal blooms, which are also known to occur in the high seas.

CHAPTER 3

BALLAST WATER MANAGEMENT SYSTEMS FOR VESSELS

In this chapter we focus on ballast water management systems (BWMS) which are currently in use as well as treatment approaches manufacturers have chosen for the development of future BWMS. The main purpose of this review is to identify the current availability of BWMS technologies worldwide. Until January 2014 we brought together information of 104 different BWMS. To achieve the ballast water discharge standards, different water treatment technologies are used, mostly in combination, and applied in different stages of the ballasting process. In general, the treatment processes can be split in three stages, i.e., pre-treatment, treatment and residual control (neutralisation). Among the 104 BWMS identified, 100 apply their treatment at the uptake, some of those BWMS require also a treatment during ballast water discharge (in-line treatment) and three BWMS apply treatment only during the voyage (in-tank treatment). The majority of BWMS use filtration or a combination of hydrocyclone and filtration as pre-treatment separation step.

The dominating treatment processes are to use an active substance, mostly generated on board by electrolysis/electrochlorination. The second most frequent treatment process is UV. BWMS to be installed for operation on vessels need to be type approved by a state. By the writing of this chapter more than 30 BWMS are type approved. It should be noted that the development of BWMS is a very dynamic market with newly proposed BWMS appearing almost on a monthly basis.

The chapter also outlines how BWMS are applied on vessels, their capacities and installation requirements, which BWMS are type approved, and what projected global market for BWMS may exist. A recent calculation on the estimated value of the global market for purchasing and installing BWMSs resulted in an estimated turn-over of possibly $50–74 billion. The chapter ends with a list of manufacturers, commercial names of their BWMS, applied treatment technologies used and links to BWMS web pages where available.

Introduction

As the entry into force of the International Convention for the Control and Management of Ships' Ballast Water and Sediments (BWM Convention; IMO 2004) is approaching rapidly the industry is more and more aware and considers ballast water management a good business. This becomes obvious when noting the high number of vessels which need to be equipped with treatment systems.

This comprehensive review of ballast water management systems (BWMS) was conducted until January 2014. In this chapter we focus on BWMS which are currently in use as well as treatment approaches the manufacturers have chosen for future BWMS.

The main purpose of this review is to identify the current availability of BWMS technologies worldwide, to briefly introduce these and their use on vessels, identify their timely availability in relation to the BWM Convention requirements, and to identify the prospects of the global BWMS market. At the beginning of this chapter the requirements that BWMS need to comply with are presented, followed by an introduction of BWMS identified, which technologies different BWMS use, how are BWMS applied on vessels, what are BWMS capacities and their installation requirements, what is the situation with BWMS testing and approvals and what is the foreseen global market for BWMS. At the end, names of manufacturers, commercial names of their BWMS, treatment technologies used and links to BWMS web pages are provided, where available.

IMO Requirements

With the Guidelines for Approval of Ballast Water Management Systems (G8) (G8 Guidelines), IMO has in 2008 adopted requirements for a comprehensive test programme to evaluate the performance and suitability of BWMS. This includes performance tests in larger scale on land under controlled conditions as well as shipboard tests to show the efficiency and seaworthiness of BWMS.

Noting some shortcomings in these test requirements some countries have developed their own requirements and test protocols, which set more stringent standards than the G8 Guidelines. One example is the USA with its Environmental Technology Verification (ETV) Program developed for the U.S. Environmental Protection Agency and the U.S. Coast Guard Shipboard Technology Evaluation Program (STEP) (NSF

International 2010; STEP 2010). At present there are many different treatment technologies available, and most of those are previously developed for municipal and other industrial applications. However, when applying those without modifications and improvements to the ballast water treatment purpose, none of these technologies have shown the capability to treat the ballast water to the level required by the BWM Convention D-2 Ballast Water Performance Standard. The setting of these proposed regulations is an important driving force for ballast water treatment technology developments worldwide. As a result, it was expected that the development and implementation of these systems will proceed at a greatly accelerated rate. However, the ambitious phase-in of the D-2 standard.

The International Maritime Organization (IMO) has adopted the International Convention for the Control and Management of Ships' Ballast Water and Sediments, in order to regulate discharges of ballast water management system (BWMS) and minimize the risk of spreading harmful aquatic organisms by ship ballast water.

Treatment of ballast water is side-Stream type which enables the system to be remotely installed away from the ballast lines. This system can be installed separately by unit. So relocation of other equipment and additional engineering can be minimized.

Key Features of Eco Guardian – Ballest's Water Management System (BMWS)

Automatic back-flushing filter unit removes large organisms and solid particles above 50 μm

- Automatic back-flushing operation starts when the pressure difference between the water inlet and outlet reaches a certain value

- The side stream of the ballasting water goes through the electrolysis unit that generates the high concentration of TRO (Total Residual Oxidants). The concentrated stream is then injected back to the main ballasting line

- Electrolysis unit (including electrolytic cells, rectifier/transformer, booster pumps, conductivity sensor and degassing devices) produces sodium hypochlorite solution which disinfects the residual plankton, pathogens, larva or spores

- Neutralization unit adds sodium thiosulfate solution into the treated ballast water to neutralize the residual TRO during de-ballasting

- When ballasting maximum TRO concentration is limited to 9.0 mg/L

- When de-ballasting MADC (Maximum Allowable Discharge Concentration) is 0.2 mg/L TRO as Cl_2

- Control system including PLC, HMI and auxiliary equipment

System General

The Ballast Water Treatments System (BWTS) can be divided into three phase :

-Ballasting

-During voyage

-De-ballasting.

Ballasting

During ballasting process, ballasting waters is first pumped through the automatic back flushing filter unit which removes large organisms and solid particles (>50μm). The filter unit is installed directly at the main ballasting line.

The side stream of the ballasting water goes through the electrochlorination unit that generates TRO of high concentration sodium hypochlorite which is then injected back to the main ballasting line. After injection, Sodium hypochlorite is mixed and diluted with main ballasting water and goes to ballast's tanks. Maximum dose concentration is 9 mg/L TRO as Cl2 which is controlled by using the on-line TRO sensor. That is, system adjusts the supplying current which determines hypochlorite production rate by monitoring TRO sensor.

Electrochlorination unit is designed to operate at the seawater salinity of 10 PSU or more. For operation in the low salinity water, it can use the seawater in the existing onboard seawater tank because this system use only small amount of seawater in comparison with the incoming ballast water flow (less than 5% of ballast's water flow).

As a side reaction of electrochlorination, hydrogen gas is generated on the cathodes of cells. Because hydrogen gas is highly explosive, it should be properly vented from the system as soon as possible whenever it is produced. Hydrogen gas is separated by hydrogen separator and is diluted less than 1% of atmospheric concentration of hydrogen by forced air blowing. Finally this diluted hydrogen gas is vented to the outside of a ship. The lower explosive limit (LEL) for hydrogen is 4% and the fourfold safety of explosion limit is applied.

During voyage

The remained organisms in the ballast tank will be disinfected by the residual effect of disinfectants.

Ballast Water Treatment Systems (BWTS) by De-ballasting

During de-ballasting process, de-ballasting water passes through only neutralization unit prior to overboard discharging. As neutralizing agent, sodium thiosulfate is injected into de-ballasting line to neutralize the residual TRO. The injection amount of sodium thiosulfate is controlled by monitoring de-ballasting flow rate and residual TRO concentration.

Two TRO sensors are used to measure the residual TRO concentration at two points, before and after the neutralization unit. TRO measurement before the neutralization unit is used to measure the residual TRO concentration and then Controller determines the injection amount of neutralizing agent into the de-ballasting line. Another TRO measurement after the neutralization unit is used to check that residual TRO has been neutralized properly.

Neutralization unit maintains Maximum Allowable Discharge Concentration (MADC) for 0.2 mg/L TRO as Cl_2. To comply with this condition, dosing rate of sodium thiosulfate is set based on the stoichiometric ratio of TRO to sodium thiosulfate at 1:2 parts. Sodium thiosulfate is injected at the inlet Ballast's Water treatment's System pump so the residual TRO in de-ballasting water is rapidly neutralized by the mixing effect at ballast pump impeller.

BALLAST WATER TREATMENT SYSTEMS

The spread of aquatic invasive species through ballast water is a major ecological and economical threat. Because of this, the International Maritime Organization (IMO) set limits to the concentrations of organisms allowed in ballast water. To meet these limits, ballast water treatment systems (BWTSs) are developed. The main techniques used for ballast water treatment are ultraviolet (UV) radiation and electrochlorination (EC). In this study, phytoplankton regrowth after treatment was followed for six BWTSs. Natural plankton communities are treated and incubated for 20 days. Growth, photosystem II efficiency and species composition are followed. The three UV systems all showed similar patterns of decrease in phytoplankton concentrations followed by regrowth. The two EC and the chlorine dioxide systems showed comparable results. However, UV- and chlorine-based treatment systems showed significantly different responses. Overall, all BWTSs reduced phytoplankton concentrations to below the IMO limits, which represents a reduced risk of aquatic invasions through ballast water.

Invasive species are one of the greatest threats to biodiversity. In the aquatic environment, there are many invasive species causing great economic and ecologic harm. Examples of this are the American comb jelly (Mnemiopsis leidyi) which is partly responsible for ecosystem shifts and the reduction of fisheries in the Caspian Sea, Sea of Azov and Black Sea. Another is the European zebra mussel (Dreissena polymorpha) which is causing fouling problems in North American lakes and rivers. Smaller organisms can also cause problems; the diatom Coscinodiscus wailesii is invasive in the North Atlantic, North Sea and Celtic Sea where it has detrimental effects on fisheries due to mucus production that clogs fishing nets.] In addition, it changes ecosystem functioning since it is indigestible to the two common herbivorous copepods and displaces native phytoplankton species. Another example is the increased spread of toxic phytoplankton blooms, such as of the dinoflagellates Alexandrium catenella or Gymnodinium catenatum which cause paralytic shellfish poisoning in humans. The most important vector for the spread of aquatic invasive species is ballast water. Because of this the International Maritime Organization (IMO) created the D-2 ballast water performance standard that set limits on the concentration of viable organisms allowed to be in ballast water at discharge. For organisms 50 micron less than 10 per m3 are allowed to be discharged. For organisms < 50 and 10 micron less than 10 per mL are allowed to be discharged. All sizes should be measured as the minimum dimension, meaning the smallest diameter of the organism.

The standard also includes three indicator microbes; toxigenic Vibrio cholera (less than 1 cfu (colony forming units) per 100 mL), Escherichia coli (less than 250 cfu per 100 mL) and intestinal Enterococci (less than 100 cfu per 100 mL). To meet these standards, a number of companies started developing ballast water treatment systems (BWTSs). These BWTSs are based on a variety of techniques, but most common are a combination of a filter to remove organisms > 50μm followed by disinfection by ultraviolet (UV) radiation or electrolytic generation of hypochlorite (electrochlorination (EC)). At the Royal Netherlands

Institute for Sea Research (NIOZ), these BWTS are tested according to IMO regulations G8 and G9.

Phytoplankton forms the basis of the marine food web and is known to survive transport in ballast water, which is why it was chosen as the focus of this book. Regrowth is defined as increase in phytoplankton concentration and viability after a BWTS treatment.

Regrowth also provides an indication of risk of introducing non-native species even after ballast water treatment according to the IMO standards.

In earlier studies, the performance of the BWTSs was measured by its effect on phytoplankton survival and regrowth during incubation experiments. Both a UV BWTS and a chemical BWTS have phytoplankton regrowth after treatment, but no statistical comparison was made.

It can be identified that several regrowing phytoplankton species after UV treatment by one specific UV-based BWTS. All regrowing species are diatoms, most notably Thalassiosira weissflogii. However, these studies did not investigate the possible differences in species responses between BWTSs. Successful tests of UV and EC treatment systems have been evaluated previously address regrowth. This is the first time that multiple systems using multiple treatment types are directly compared.

This study compared regrowth in six BWTSs. Three BWTSs used UV, but differ in the number of UV reactors and type of UV source used.

Two BWTSs used EC, generating hypochlorite by electrolysing seawater, one system generated chlorine dioxide (CD) by adding sulphuric acid (H_2SO_4) and a mixture of sodium chlorite ($NaClO_2$) and hydrogen peroxide (H_2O_2) together.

The six systems tested in this paper can be grouped into two general categories, disinfection by UV radiation and disinfection by chlorine. The main research question was: is there a difference in performance between these two types of ballast water treatment?

In order to answer this question first a comparison was made between the BWTSs that use the same method to see if different systems using similar methods also have similar performance.

Additionally, for both UV radiation and chlorine chemistry a dosage experiment was performed to investigate the effects that an increased or decreased dosage had on the organisms in ballast water. Finally, the performance of all systems was compared. Phytoplankton performance was investigated by following cell concentration, Photosystem II (PSII) efficiency (as indicator of physiological status of the photosynthetic machinery) and species composition during and after BWTS treatments.

CHAPTER 4

BALLAST WATER TREATMENT SYSTEMS

HYPOTHETIC SCENERIOS AND OBSERVATION

Observation According to images below The UV1 BWTS used a 50 µm disk filter and one UV reactor with medium pressure (broad wavelength) UV. The UV2 BWTS used a pre-filtration over a 200 µm mesh filter, followed by a 50 µm mesh filter and two UV reactors with low pressure (254 nm) UV. The UV3 BWTS used a 20 µm mesh filter and three UV reactors with low pressure (254 nm) UV radiation. The CD treatment system used a 40 µm mesh filter followed by an addition of chlorine dioxide. The EC1 system used a 40 µm mesh filter and electrolytic chlorination to generate hypochlorite, which is neutralized on discharge using sodium bisulphite.

The EC2 system used a 200 µm filter, a cyclone to separate particles down to 20 µm and electrolytic chlorination to generate hypochlorite, which is neutralized on discharge using sodium bisulphite. The tests for UV1, UV2, CD, EC1 and EC2 consisted of filling two 250 m3 simulated ballast water tanks (one treated and one control for each treatment system) at a speed of 200 m3 per hour. Water was pumped up from the NIOZ harbour, passed through the pump and the treatment system after which intake samples are taken.

For the control tank, water also went through the pump but by-passed the entire treatment system. Thus control samples are not filtered. Intake samples of the controls are taken after the pump. The tests for UV3 are performed with three tanks of 250 m3, one control and two treated. Both control and treated water was kept in the simulated ballast tanks for 5 days (as described in the IMO guidelines). After this 5-day period, the water was discharged. All three UV BWTSs applied a second UV treatment at discharge. Water from the treated tanks thus passed again through pump and treatment system, after which discharge samples are taken. Both EC BWTSs added a neutralizing agent on discharge.

Table 1. Overview of treatment details for all systems.

System	Pre-treatment
UV1	50 μm disc filter
UV2	200 μm mesh filter and 50 μm mesh filter
UV3	20 μm mesh filter
CD	40 μm mesh filter
EC1	40 μm mesh filter
EC2	200 μm filter and hydrocyclone

Treatment

UV1	One UV reactor, medium pressure. Treatment at both intake and discharge
UV2	Two UV reactors, low pressure. Treatment at both intake & discharge
UV3	Three UV reactors, low pressure. Treatment at both intake & discharge

CD	Chlorine dioxide addition through mixing of two chemicals
EC1	Hypochlorite addition through electrolysis, neutralized by sodium bisulphite on discharge
EC2	Hypochlorite addition through electrolysis, neutralized by sodium bisulphite on discharge

Results

Comparison of BWTSs using UV radiation

There are common patterns between the three UV radiation based BWTSs. During the first days after treatment phytoplankton concentrations only gradually decreased, both at intake (single UV treatment) and discharge (second UV treatment). Phytoplankton concentrations in control samples stayed relatively stable while the PSII efficiency went down over the 20-day incubation period. In treated samples the PSII efficiency was close to zero during the first days, but increased after 4–8 days to a peak value after which it decreased.

All UV BWTSs showed regrowth after treatment, with the exception of the UV2 BWTS where there was no regrowth after the second UV treatment. Regrowth occurred in 6–10 days after the first UV treatment and in 7–12 days after the second UV treatment. Phytoplankton concentrations after regrowth are sometimes much higher than initial control concentrations.

Genetic analysis of regrowing phytoplankton species was only performed on the UV2 and UV3 BWTSs. Always only one species was found in the control sample, but this is a known limitation of the analytical method.

The species found in the control sample did not match any of the species found after treatment for any of the experiments. While Thalassiosira pseudonana survived and re-grew after treatment with the UV2 system, this species apparently did not survive treatment with the UV3 system where it was detected in the control, but after treatment only Thalassiosira weisflogii was found.

UV dose response experiment

The UV dose response experiment of the UV2 BWTS showed reduced phytoplankton concentrations with increasing UV dose. The control showed a gradual decrease in phytoplankton concentration. The gradual decrease in phytoplankton concentrations, which was observed during normal incubation experiments, was visible at all dosages; at higher dosages the decrease was more pronounced. Performance at a dosage of 75% was similar to performance at a dosage of 100%, but at 50% the drop in phytoplankton concentration was much less. All UV doses immediately reduced PSII efficiency to below 0.1, except for the 25% dosage. PSII recovery occurred fastest at 50% and slowest at 400%. PSII efficiency in the control sample showed a gradual decrease while in treated samples it showed a peak after regrowth after which it decreased.

Comparison of chlorine-based treatment systems

The CD BWTS reduced phytoplankton concentrations and PSII efficiency immediately. This decrease continued during the first three incubation days, while phytoplankton concentrations in the control stayed relatively constant. Of the three incubation experiments with the CD

BWTS regrowth was observed twice, once at T12 and once at T20. Phytoplankton concentrations.

EC dosage experiment

Phytoplankton concentrations in the control showed a strong peak followed by a reduction to below the starting concentration, after which the phytoplankton concentration remained stable. PSII efficiency showed a similar pattern, starting high but afterwards decreasing to a low but stable level. Without neutralization

Genetic identification of phytoplankton species was performed on the dosage experiment of the EC2 BWT. None of the species detected in the control sample are found after treatment but neither are the species detected after treatment found in the control. Navicula phyllepta and Chaetoceros socialis are found both in samples with excess hypochlorite and excess sodium bisulphite. Emiliania huxleyii was only found in samples with excess sodium bisulphite.

Comparison of UV and EC treatment systems

The NMS analysis of the comparison between all treatment types revealed a difference between treatments (R 0.33, P < .01). Two different groups are found: the first group consisted of 1 UV and the second group of EC, EC BS, CD and 2 UV (Figure 7). However, 1 UV can be considered an incomplete treatment, since in all UV systems water is also treated on discharge. When an ANOSIM analysis was conducted excluding the 1 UV data, the result was different. A significant difference was found between 2 UV and CD (P < .05) and between

2. UV and EC, including EC BS (P < .05). No significant difference was found between CD and EC.

Comparison of UV BTWS

Despite their differences in configuration (both filters and reactors) and UV wavelength, the three UV radiation-based treatment systems produced similar results. All of them showed a gradual decrease in

phytoplankton numbers after treatment. This 'delayed effect' emphasizes the importance of PSII efficiency measurements for phytoplankton, since phytoplankton concentrations (as measured by flow cytometry) are higher than IMO standards immediately after treatment.

However, the phytoplankton has very low PSII efficiency and disintegrates over time, reaching concentrations below the IMO standards. Regrowth occurred in all systems, and in all systems around the same time, between 6 and 12 days. On discharge initial phytoplankton concentrations are lower than at intake and the 'minimum values' for phytoplankton concentrations (as shown in Figure 1) are also lower. The results of genetic identification of phytoplankton are also similar for the two UV BWTSs tested (UV2 and UV3).

For both systems, Skeletonema costatum was a regrowing species, but the most frequent regrower are species belonging to the genus Thalassiosira. After treatment with the UV2 BWTS, Thalassiosira pseudonana was the main regrower while after treatment with the UV3 BWTS it was Thalassiosira weissflogii.

This indicates that Thalassiosira is more resistant to UV radiation than other phytoplankton genera, matching the scholars. It is interesting however that Thalassiosira pseudonana was found in the control samples for the UV3 system, but after treatment Thalassiosira weissflogii was the regrowing species. It is unknown if this is really a species shift or a misidentification by the genetic analysis method.

UV dosage

The UV dosage experiment of the UV2 BWTS showed that even with treatment dosage reduced to 75% (by disconnecting half the lamps of the second UV reactor), the phytoplankton concentration and viability behaved similar to 100% treatment. This suggests that even when treatment effectiveness is reduced the system will still perform up to IMO standards.

Additionally, 200% and 400% dosage (achieved by lowering flow speed so water spent more time in the reactors) showed a stronger reduction in phytoplankton concentration than the 100% treatment, but regrowth occurred around the same time.

This suggests that an increase in UV radiation dose will not eliminate the possibility of regrowth.

It should be noted that all UV doses in the UV-dosage experiment are from a single treatment; in normal UV ballast water treatment the water would get a second treatment before discharge which would further lower organism numbers and possibly delay regrowth.

Comparison of CD and EC BWTS

Both EC BWTSs and the CD BWTS showed an immediate decrease in phytoplankton concentrations after addition of chemicals. The main difference of the CD BWTS compared to the EC BWTSs is that the CD BWTS does not add a neutralizing agent upon discharge.

For both CD and EC without neutralization regrowth occurred in part of the experiments (regrowth in two out of three experiments for CD and one out of three experiments for EC) (Table 4). In two of the three regrowing experiments, regrowth only occurred just before the end of the experiment (Day 20 for the third CD experiment and Day 19 for the first EC1 experiment) (Table 4). It is therefore recommended to conduct regrowth studies on this type of system for longer than 20 days.

The EC results with neutralization are very different; all three showed regrowth suggesting that

EC Dosage

Without neutralization no regrowth occurred, but even partial neutralization of the hypochlorite with sodium bisulphite resulted in regrowth within 20 days. Regrowth also occurred when excess sodium bisulphite was added. Neutralization is therefore an important part of making the treated ballast water safe for discharge since even partial neutralization seems to mitigate the harmful effect of hypochlorite. While the major regrower from the UV experiments, Thalassiosira weissflogii, was present in the control samples, it was not detected after treatment. The regrowing species for the EC experiment are Navicula phyllepta, Chaetoceros socialis and Emiliania huxleyii of which Emiliania huxleyii only occurred in samples that are completely neutralized or had an excess of neutralizing agent.

Comparison of UV and CD/EC BWTS

As expected, statistical analysis showed that there was a difference between treatment types. What was not expected was that 2 UV grouped together with EC and CD, with 1 UV grouping separately. With 1 UV excluded from the analysis, the data showed the expected pattern with no significant difference between CD and EC but significant difference between UV and chemical treatments. The most important contributing factors to this difference are the lower initial reduction in phytoplankton concentrations, more gradual slope of decrease and earlier start of regrowth of the UV systems.

In both UV and CD/EC BWTSs, phytoplankton concentrations after regrowth are higher than phytoplankton concentrations in the control. This is probably due to the fact that most regrowing species are small (10 micron or smaller) while the control consisted of a mixture of NMS of all incubation experiment data. Two main groups are significantly different, 1 UV (light grey triangles) and

2 UV, CD, EC and EC BS (light grey squares). There are also three outliers. (The EC outlier is the EC experiment with the earliest regrowth (EC2, EC BS.) The 2 UV outlier is the only UV experiment with no regrowth (UV2 experiment 2), set to regrowth on Day 20. The 1 UV outlier is an experiment with a low initial reduction, high minimum value and high slope of decrease (UV1, experiment 2).

All BWTSs compared used a filtration step, but the mesh size of these filters varied between 20 and 200 micron. However, when comparing the results of UV2 (50 micron filter mesh) and UV3 (20 micron filter mesh) regrowing species are similar. Regrowing species of the EC2 BWTS, which used a 200 micron filter mesh, are all below 10 micron in minimum dimension. Since all regrowing species are smaller than the smallest filter mesh used, this implies that the size of the filter mesh did not affect the regrowing species.

Regrowing phytoplankton species differed between UV and chemical systems. This indicates that there is not one 'super plankton' resistant to all treatments. Different ballast water treatment techniques have different challenge species, whether it is because of a built-in resistance or because of a life history which allows it to escape the effects of the treatment.

When comparing UV and EC BWTSs from a ship owner perspective, UV systems have the advantage that no chemicals need to be carried aboard. In case of emergency, the ballast water can be discharged at any point without environmental problems. As a disadvantage, when scaling up the system both extra filtration and UV units need to be installed, and this requires more space.

Additionally, UV reactors are large energy-consumers, especially in low UV-transmittance waters where the ballast water flow rate might even have to be reduced in order to treat the ballast water with the required minimum dose. EC systems are easier to scale up since the reactor requires only minimal scaling; only extra filtration units are required. However, this system needs to have chemicals onboard. In the case of chlorine dioxide, it is the two components of the reactor mixture and with EC systems it is sodium bisulphite to neutralize the hypochlorite before discharge. Additionally, EC systems require salt water to produce the active substance. When operating in fresh water, a supply of salt water will have to be carried onboard.

All six systems described in this paper met the IMO D-2 standards for BWTSs. The IMO standards do not ask for ballast water free of organisms, but set a strict and low maximum standard. As the present experiments have shown, organisms can regrow after treatment by each of the six BWTSs when provided with optimal growth conditions. This means that the risk of invasive species is not eliminated by ballast water treatment. On the other hand, the concentration of organisms introduced is strongly reduced, which results in a reduction in propagule pressure. Propagule pressure is a key factor in the success of non-native species in a new environment. Even though ballast water treatment is not 100% effective, it still greatly reduces the threat of invasive species spread through ballast water.

The UV1 BWTS used a 50 μm disk filter and one UV reactor with medium pressure (broad wavelength) UV. The UV2 BWTS used a pre-filtration over a 200 μm mesh filter, followed by a 50 μm mesh filter and

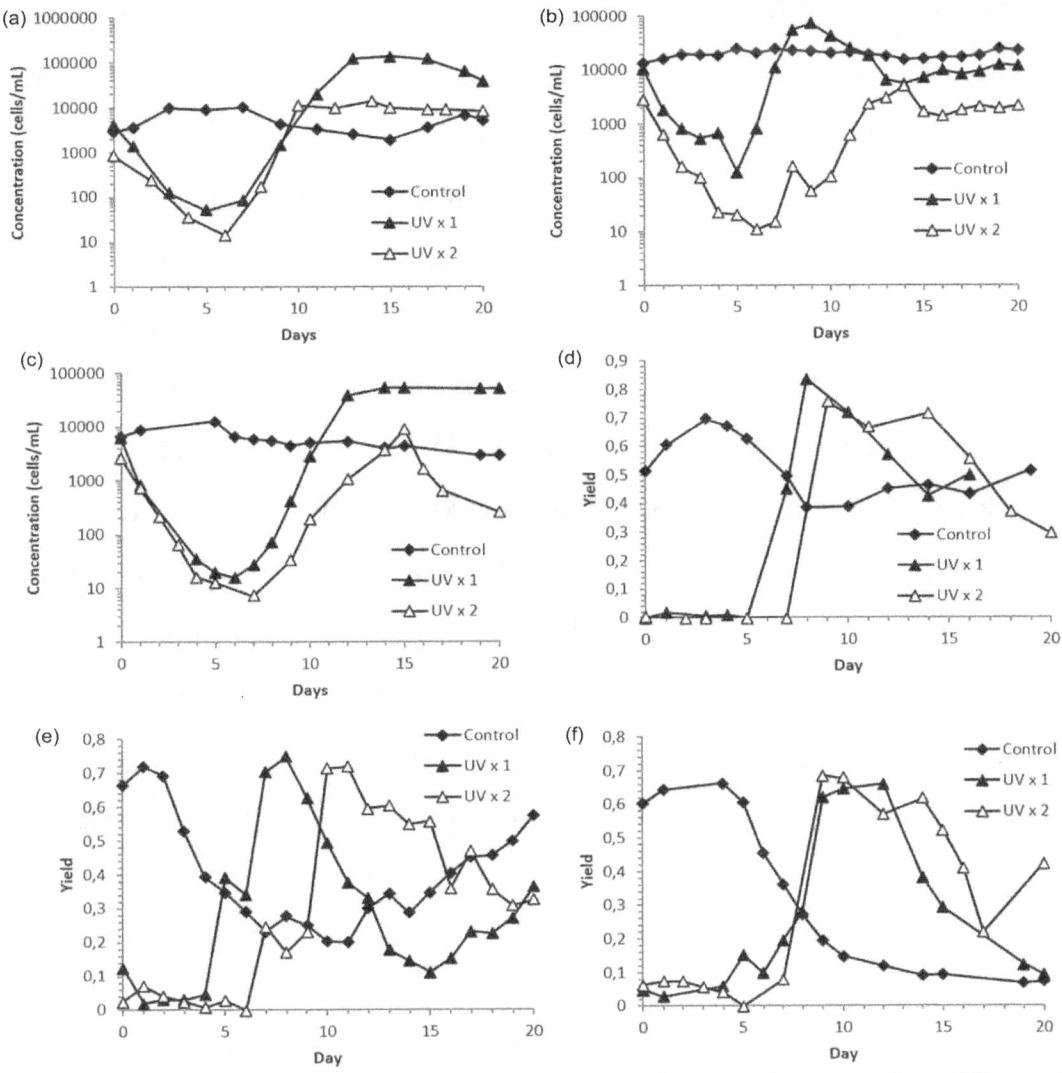

two UV reactors with low pressure (254 nm) UV. The UV3 BWTS used a 20 μm mesh filter and three UV reactors with low pressure (254 nm) UV radiation. The CD treatment system used a 40 μm mesh filter followed by an addition of chlorine dioxide. The EC1 system used a 40 μm mesh filter and electrolytic chlorination to generate hypochlorite, which is neutralized on discharge using sodium bisulphite. The EC2 system used a 200 μm filter, a cyclone to separate particles down to 20 μm and electrolytic chlorination to generate hypochlorite, which is neutralized on discharge using sodium bisulphite. The tests for UV1, UV2, CD, EC1 and EC2 consisted of filling two 250 m3 simulated ballast water tanks (one treated and one control for each treatment system) at a speed of 200 m3 per hour. Water was pumped up from the NIOZ harbour, passed through the pump and the treatment system after which intake samples are taken. For the control tank, water also went through the pump but by-passed the entire treatment system. Thus control samples are not filtered. Intake samples of the controls are taken after the pump. The tests for UV3 are performed with three tanks of 250 m3, one control and two treated. Both control and treated water was kept in the simulated ballast tanks for 5 days (as described in the IMO guidelines). After this 5-day period, the water was discharged. All three UV BWTSs applied a second UV treatment at discharge. Water from the treated tanks thus passed again through pump and treatment system, after which discharge samples are taken. Both EC BWTSs added a neutralizing agent on discharge.

Protecting Aquatic Ecosystems

To protect aquatic ecosystems and human health and to reduce economical expenses from the impact of invasive species, international guidelines on ballast water management are developed in the 1990s eventually resulting in The International Convention for the Control and Management of Ships' Ballast Water and Sediments (BWM Convention) adopted by the International Maritime Organisation (IMO) in 2004 (IMO, 2004) and which entered into force in 2017.

The convention states that all ships must manage their ballast water using ballast water management systems (BWMS). The BWMS must be type approved in accordance with the guidelines G-8, and guidelines G-9 if the BWMS is using active substances (IMO, 2008, IMO, 2016a, 2016b).

The performance of BWMS must comply with set discharge standards related to the number of viable organisms in defined size-classes. For the 10–50 µm size-class, which mainly consists of phytoplankton, the discharge standard is < 10 organisms ml^{-1} . BWMS use different methods to kill organisms in ballast water: Some systems use biocidal compounds and others treat water using electrolysis or UV irradiation combined with a physical solid-liquid separation process such as filtration. UV treatment systems have already been installed in large numbers and it is estimated that > 50% of the BWMS will be based on this technology in the future (mpnballastwaterfacts.com, 2017; www.imo.org, 2017).

Two types of UV treatment can be applied:

A monochromatic system approach with low pressure (LP) mercury lamps emitting UV irradiation within the UVC (germicidal) range (100–280 nm). Approximately 85% of the power output is concentrated at 253.7 nm where it specifically affects the integrity of DNA and RNA in the cells that show maximum absorption at around 260 nm. In a polychromatic system approach, medium pressure (MP) lamps emit UV light at a higher total intensity with the energy distributed at a broader spectrum.

DNA is still affected at 253.7 nm, but at a lesser degree compared to LP UV systems (about 2.7% of the total energy between 200 and 300 nm was emitted at 253.7 nm in our system). However, a broader energy spectrum in MP.

To protect aquatic ecosystems and human health and to reduce economical expenses from the impact of invasive species, international guidelines on ballast water management were developed in the 1990s eventually resulting in The International Convention for the Control and Management of Ships' Ballast Water and Sediments (BWM Convention) adopted by the International Maritime Organisation (IMO) in 2004 (IMO, 2004) and which entered into force in 2017.

The convention states that all ships must manage their ballast water using ballast water management systems (BWMS). The BWMS must be type approved in accordance with the guidelines G-8, and guidelines G-9 if the BWMS is using active substances (IMO, 2008, IMO, 2016a, 2016b). The performance of BWMS must comply with set discharge standards related to the number of viable organisms in defined size-classes. For the 10–50 µm size-class, which mainly consists of phytoplankton, the discharge standard is < 10 organisms ml−1.

BWMS use different methods to kill organisms in ballast water: Some systems use biocidal compounds and others treat water using electrolysis or UV irradiation combined with a physical solid-liquid separation process such as filtration. UV treatment systems have already been installed in large numbers and it is estimated that > 50% of the BWMS will be based on this technology in the future.

Advanced Water Treatment Systems for Large Ships

Ships carrying cargo, cruise ships, and tankers all carry ballast water that is stored from one point of call and then usually discharged at another point of call. Although this is a necessary process and cannot be stopped, there is a hazard to the environment. When ships take on water in one area and then put it back into the water in another area, harmful plants, animals, bacteria, and viruses are spread all over the globe. This is the reason advanced water treatment systems are needed and especially a ballast water treatment system.

There are several factors that need to be considered when choosing a water treatment system. Although there is no system that will effectively remove all types of hazards from ballast water, there are several systems that will help create a better aquatic environment. When choosing a system consider the following:

- Safety of the passengers and crew

- Expense of technology; installing and running

- Ease of using and maintaining the system

- Effectiveness of removing harmful organisms

- Interference with normal travel and operation of the ship

Once all of these items have been considered, finding a water treatment system that fits the unique needs of each ship is the next step. For many captains, or owners of a ship that carries ballast water, the expense of operation is high on the priority list. No one wants to spend a lot of money or time installing or maintaining a new system. When searching for a water treatment system, find one that is affordable, easy to use, easy to maintain, and will meet the requirements of marine safety. Once the system is installed, it should not require a lot of man hours, fuel consumption, or maintenance to keep it running effectively and completing the required job.

Although most ships currently use a ballast exchange system which involves the ship exchanging water in the middle of the ocean, there are new regulations which could make this a costly endeavor. There are too many variables to consider with this system that makes it unreliable. Often, not all of the ballast water is exchanged and organisms and sediment are carried into another area. New tests have been developed to gauge whether the water has been exchanged in an unauthorized location. For these reasons, an advanced water treatment system will save money in the end.

Overall, installing a ballast water treatment system will be good for the environment and make sure invasive organisms are not introduced into nonnative areas. Once all factors have been considered, a retrofitted system can be installed and begin working immediately. With just a little training on the operation and upkeep of the system, ships carrying ballast water will no longer cause a danger to humans or marine life.

Why is ballast water treatment needed?

Ballast water taken on in one ecological zone and released into another can introduce invasive and nuisance aquatic species that may have detrimental impacts on the biodiversity, economy or human health of the receiving community and may in time become a serious threat to the environment.

Bio-invasion is one of the four greatest threats to the world's oceans today, alongside land-based sources of marine pollution, the over exploitation of living marine resources and the physical alteration and destruction of marine habitats.

More than 90 per cent of global trade is carried by sea, and each year transfers of up to 12 billion tonnes of ballast water take place around the world. In order to reduce the incidence of bio-invasions, ballast water treatment reduces or renders inactive 99.9% of the living organisms in the ballast water.

The Difference Between Ballast Water Exchange And Ballast Water Treatment

This FAQ is aimed at the crew onboard a ship that has ballast water tanks and only addresses, in a birds view, the operational differences and the main technical differences. The biggest difference is that Ballast Water Exchange is done in mid ocean, where Ballast Water Treatment is done in port during the cargo handling operations.

Ballast Water Exchange

Ballast Water Exchange is a very demanding procedure which required a tedious preparation because each exchange method has its own pros and cons. All procedures require continuous attention to keep the vessel and her crew safe. Points of attention are keeping the propellers submerged, keeping the hull stress within limits, maintaining the visibility for safe navigation, maintaining the (intact) stability, preventing possible slamming of the bow, etc.

Most of the crew sailing on ships with ballast water tanks already use a Ballast Water Exchange (BWE) method to comply with regulations. A quick recap learns us that Ballast Water Exchange is a method where coastal or port water is replaced by mid-ocean water during the voyage. Ballast Water Exchange methods in use are known as:

• Sequential exchange, where the ballast water tanks are made empty and filled with mid ocean water;

• Flow through exchange, where the mid ocean water, in three complete cycles, is pumped through in the tanks while the coastal ballast water leaves the tank simultaneously through the overflow;

• Dilution method, where the mid ocean water, in three complete cycles, is pumped in through the top of the ballast water tank, and where the coastal ballast water leaves the tank simultaneously through the bottom connection of the tank.

The whole process and requirements are laid down in a Ballast Water Treatment Plan where the execution of the Ballast Water Exchange procedures are recorded in a Ballast Water Record Book which is to be kept up-to-date and available for review and survey by the port authorities or there representative.

Ballast Water Treatment

BALLAST WATER TREATMENT SYSTEMS

Ballast Water Treatment is often, when started up, a fully automatic procedure which have to be monitored for correct operation. When Ballast Water Treatment is applied, there is no more need to apply the Ballast Water Exchange procedures. Of course Ballast Water Exchange can be used, after approval for contingency operation. Many vessels are in the process of or have already been retrofitted with a ballast water treatment system. A typical retrofit installs a ballast water treatment in the existing ballast water system.

Depending on the system in use, ballast water treatment is required at ballast water uptake and during ballast water discharge. Damen Green has partnered with three suppliers, two are specialized in continuous filtration and UV disinfection and one has specialized in continuous filtration with electro-chlorination. The electro-chlorination system is derived from the well-known Chloropac system for seawater cooling water treatment.

When continuous filtration and UV disinfection, in an automatic and monitored process, is applied; during ballast water uptake, the ballast water is pumped by the own ballast pump through the continuous filter stage followed by UV-disinfection into the selected ballast water tank(s) as usual. During the uptake the filter is automatically cleaned (backwash) while the uptake procedure keeps ongoing. The backwash is pumped overboard by the backwash pump which is part of the ballast water treatment installation. During ballast water discharge, in an automatic and monitored process, the ballast water from the ballast water tanks are pumped with the own ballast water pump through the UV disinfection stage only, overboard as usual.

CONCLUSION

When continuous filtration with electro-chlorination, in an automatic and monitored process, is applied; during ballast water uptake, the ballast water is pumped by the own ballast pump through the continuous filter stage (main stream) disinfection into the selected ballast tank(s) as usual. Directly after the ballast pump a water sample is taken to determine the amount of Sodium Hypochlorite which needs to be added to achieve the required level of disinfection.

The Sodium Hypochlorite is injected in the main stream before the ballast water enters the selected ballast water tank(s). The Sodium Hypochlorite is, in a side stream, made onboard from seawater with the use of an electrolyzer and a degassing module to dilute and release the electrolyzer exhaust gases which are a residue of the Sodium Hypochlorite production.

During ballast water discharge an analyzer connected to the ballast water piping before the overboard sea chest, determines the amount of Sodium Hypochlorite left in the ballast water to discharge. If necessary, residual chlorine is automatically neutralized with Sodium Sulphite before it is discharged overboard as usual, if no chlorine is measured, the ballast water is pumped, without further treatment, overboard as usual.

www.ingramcontent.com/pod-product-compliance
Lightning Source LLC
Chambersburg PA
CBHW080621220526
45466CB00010B/3420